MODEL VIEW CULTURE

New writing on technology, diversity and culture.

2015 Quarterly No. 2

San Francisco, California

June 2015

Model View Culture is published by
Feminist Technology Collective, Inc.

SIL Open Font Licensed Typography by:

Lato by Łukasz Dziedzic, latofonts.com

Libre Baskerville by Pablo Impallari, Rodrigo Fuenzalida, impallari.com

Abril Fatface by Veronika Burian & José Scaglione, type-together.com

Printed by 1984 in Oakland, California USA

This Quarterly is offset printed in the USA by a woman-owned small business, on recycled paper, using zero-VOC and recycled inks, with animal-free bookbinding.

ISBN: 978-0-9915166-5-0

To order directly from the publishers, please visit:
http://modelviewculture.com

Contents

Letter from the Editor

Welcome to the second issue of Model View Culture's 2015 subscription!

We are officially a year and half old! Feels like both way longer and way shorter, depending on the day. :)

Our second year has been fantastic in many ways so far. Over time, we've built a more mature platform, which has been incredibly rewarding – we've been meeting and publishing many new authors, and expanding our coverage area to include more creative work as well as subjects we haven't delved into before. And, our audience is growing, thanks to the support of readers like you.

This is great news for a publication in its sophomore year!

With about 18 months of publishing behind us, and our sixth Quarterly edition out now, we've improved many of our core processes, with a better fundamental mastery over the business in everything from editorial to finances. While our first year was focused on getting up and running and gaining initial financial stability, I see our second year as being about getting into a more efficient rhythm, developing our editorial platform, and reaching a larger audience with new stories, authors and topics.

At the same time, this year hasn't been without its challenges. For one, the growth of online hate groups targeting the tech and gaming communities has been incredibly worrying, and a real obstacle to our growth on several levels. Instead of investing in some of our major goals for 2015 – like hiring – we've been saving up an emergency online harassment and legal fund so we're prepared in the case of a major attack on MVC. After being personally targeted by these groups earlier this year, I'm taking the risk factors incredibly seriously. As a business owner, it's been indescribably frustrating to dedicate so much time and money to defending against this type of terrorism, instead of focusing on our readers and the growth of our business. On a personal level, it's taken a toll on my mental health and energy as an editor. I have grave concerns about the impact of these groups on not only our publication, but the growth of diversity in tech efforts overall. However, we stay hopeful that the tech community will better work together to combat these threats at both a practical and cultural level this year and in years to come.

In other news, you might have heard our print shop, 1984 Printing, suffered a severe fire earlier this year. However, we're really happy they've gotten back up and running, and we really appreciate all of the Model View Culture readers who helped support their crowdfunding campaign either by donating or spreading the word! Part of what we love about our work is getting to collaborate with other local and woman-owned small businesses, and 1984 has been with us from the start as a huge part of our family – so a big shout-out to them and the amazing work they do for us and other independent publishers, writers and artists. Thanks 1984!

So, that's the inside scoop on goings-on at MVC! We really hope you enjoy this issue, and encourage you to share it with your friends and colleagues once you're done. Pass it on!

Shanley Kane

Shanley Kane
CEO, Model View Culture

The Core Belief Keeping Marginalized Groups Out of Tech

Dr. Nicole Forsgren and Jez Humble

"Sometimes the situation is only a problem because it is looked at in a certain way. Looked at in another way, the right course of action may be so obvious that the problem no longer exists."
- Edward de Bono

We hear constant complaints about the "talent shortage" from our peers in the tech industry. The refrain is sounded on stages, and in news articles and think pieces across the field; big tech is throwing millions of dollars at the "pipeline problem," and the shortage is considered so urgent that many industry leaders have lobbied Congress to extend more visas to skilled foreigners. However, this supposed dearth of talent could be solved at a stroke simply by looking for it where we aren't used to looking: within the large population of marginalized people tech has been keeping out.

Women and people of color are strongly underrepresented in technical roles in the US software industry. The best available data shows that women make up about

15% of technical staff.[1] The Bureau of Labor Statistics has more optimistic numbers for women in computer and information management, at 27%, but reports only slightly more than 5% Black and 5%xs Latino workers in those ranks; a 2012 AFL-CIO study also reports Black and Latino workers in math and computing occupations at 6% - 7%.[2] We don't have good data on the exit of minority populations from tech, but we know that women drop out of the technology industry at a much higher rate than men: after 10 years, 41% of women will leave their technology careers compared to 17% of men.[3] This is not because they are dropping out of the workforce, but because they are switching to non-STEM careers.[4]

We know that differences in ability aren't behind this underrepresentation in STEM: "biological sex differences in inherent aptitude for math and science are small or nonexistent,"[5] and the same holds true for differences across racial and ethnic groups. Sadly, research suggests it is *the beliefs of professionals in the industry* keeping marginalized people from joining tech, and driving out a disproportionate number of those who do.

Beliefs or Abilities: Which is More Important?

In particular, two beliefs conspire to drive under-represented groups from the field: the belief that innate ability and brilliance are required to succeed; and the belief that certain groups of people do not have that innate bril-

1 "Women Technologists Count: Recommendations and Best Practices to Retain Women in Computing," Anita Borg Institute
2 "How Tech Stays White," Von Diaz and Jamilah King, Colorlines
3 "Women In IT, The Facts," NCWIT
4 "Why Do Women Leave Science and Engineering?" Jennifer Hunt
5 "Science Faculty's Subtle Gender Biases Favor Male Students," Moss-Racusin, et al.

liance.

These beliefs are pervasive in technology. In fact, the myth that certain software engineers are significantly better than others -- the "10x developer" myth -- is persistent, and it is widely assumed that these people are born and not made. For example, Paul Graham makes this assumption explicit in his article urging higher levels of immigration, "Let The Other 95% In":

"What the anti-immigration people don't understand is that there is a huge variation in ability between competent programmers and exceptional ones, and while you can train people to be competent, you can't train them to be exceptional. Exceptional programmers have an aptitude for and interest in programming that is not merely the product of training."

Unfortunately, beliefs in "innate ability" have the power to not only influence VC investing patterns -- and potentially immigration policy -- but also to keep minorities out of tech.

Beliefs about innate ability and their effects on gender imbalance in STEM professions were investigated by Dr. Sarah-Jane Leslie and her collaborators, and published in Science magazine in January 2015. This large-scale, nation-wide study examined the beliefs and work practices of STEM PhDs from multiple fields of study, and how these beliefs and practices correlated with field demographics. The study tested:

1) gender differences in willingness or ability to work long hours;

2) gender differences in the high end of the aptitude distribution (in other words, ability);

3) the extent to which a particular discipline requires systemizing (the need to understand abstract principles and rules) over empathizing (the need to under-

stand human thoughts and feelings); and

4) the extent to which practitioners *believe* that raw, innate talent is the main requirement for success.

The authors found that across STEM fields, only the final hypothesis correlated in a statistically significant way with the percentage of US PhDs who were female.

In other words, there were *no significant differences between men and women* in the willingness or ability to work long hours, no significant differences between men and women in aptitude, and no significant gender differences around the systematic vs. empathetic nature of the field. The only significant differences were in the *beliefs* that innate talent was required for success. In other words, fields where professionals believed innate ability and raw talent were required for success had significantly less women. These fields? Math, physics, computer science, and engineering.

Shame on us. It is simply our *belief* that women lack the innate intellectual ability to do the work that is keeping them out of certain STEM fields. The research then doubles down and reports that African Americans are poorly represented for precisely the same reason -- the pervasive belief that they lack the innate talent to do the work.

Unfortunately this belief also sneaks into our education system, as shown by a study from 2012[6] in which researchers asked 127 biology, chemistry, and physics professors from across the USA to evaluate a science laboratory manager who had just applied for the position in their lab. The professors were all given identical job application materials from an undergraduate science student, but the application packets were randomly as-

6 "Science Faculty's Subtle Gender Biases Favor Male Students," Moss-Racusin, et al.

signed either a male or female-coded name. Participants were asked to rate the student's "competence and hireability, as well as the amount of salary and the amount of mentoring they would offer the student." Applicants with male names were perceived as significantly more competent and hirable, and deserving of higher pay and mentorship. Perhaps most interestingly, both male and female professors demonstrated the same bias, showing that it is not intentional or explicit but rather "shaped by implicit or unintended biases, stemming from repeated exposure to pervasive cultural stereotypes that portray women as less competent." Other studies have shown the same effects in different domains, as well as a similar effect with regard to race.[7]

What Can We Do?

The results of the studies are clear: pervasive cultural stereotypes that "brilliance" is required to succeed in programming, and that women and people of color are less competent, are the greatest barriers to increasing and maintaining representation. There are several important steps we can take to correct these problems.

First, industry leaders and managers at every level must actively and consistently work to combat these stereotypes. We must stop talking about "10x developers," ninjas, rock-stars, and the scarcity of "talent." The idea that you create great organizations by hiring brilliant people has been roundly debunked (for example, Enron was explicitly managed according to this principle.) As Malcolm Gladwell points out in his article on the "Talent Myth," which expands on Enron's collapse, it is not individuals

---

7 See, for example, "Are Emily and Greg More Employable Than Lakisha and Jamal? A Field Experiment on Labor Market Discrimination," Bertrand, M. and Mullainathan, S.; and "Study Reveals Ethnic Discrimination in Recruitment of Young Workers," International Labour Organization.

that make organizations smart, but teams:

"...our lives are so obviously enriched by individual brilliance. Groups don't write great novels, and a committee didn't come up with the theory of relativity. But companies work by different rules. They don't just create; they execute and compete and coordinate the efforts of many different people, and the organizations that are most successful at that task are the ones where the system **is** *the star."*

Second, we must look beyond ideas of innate ability to focus on what really leads to success in the tech world. Neither software engineers nor IT operations personnel have gotten where they are simply through inherent brilliance, formal education, or training. Rather, as with sports or music, mastery is gained through repeated practice. This is highlighted in the NPR *Planet Money* episode "When Women Stopped Coding" by Steve Henn, which examined the steep drop-off in women's graduation from computer science degrees that occurred in the 80s. Henn found this exodus of women from CS degrees was correlated with the rise of cheap home computers, which were targeted almost exclusively at boys. Students who spent time programming on home computers experienced a significantly easier time adjusting to the course material, causing many women -- who had not had exposure to home computers -- to drop out. Some universities have now redesigned their CS courses to ensure they compensate for effects like this: UC Berkeley's CS 101 course, "The Beauty and Joy of Computing", enrolled 106 women and 104 men in its first year. Similarly, the industry must look for innovative ways to engage with communities that haven't traditionally been in tech, because it is these communities that may bring us the best insights and opportunities for creativity and innovation.

Third, companies should review their processes and correct obvious inequalities. Review staff by role and look for statistically significant differences in salary when people are aggregated by gender and by race, then correct for them. Correct any outliers. Monitor tenure, rate of advancement and job satisfaction by gender and race, make the results visible, and update them. Examine recruiting and performance reviews for evidence of bias -- are women more likely than men to be rejected in interviews for "culture fit", or more likely to receive negative performance reviews? Do recruitment specs use language that is likely to put off women? Hire an external expert to review interactions, policies, and HR processes, make recommendations, and return regularly to monitor implementation and review progress.

Finally, companies must invest in developing their people and creating an environment where people can practice and improve. We can't over-emphasize the extent to which many of the problems in the software industry today are the result of focusing on short-term goals and hiring the necessary talent to achieve them, rather than focusing on longer-term goals and developing our people in pursuit of them. Managers should help software developers set stretch goals for their development, provide them with the support, training and slack time to gain the necessary skills to meet these goals, and ensure the work they are doing provides opportunities to work towards them.

Fixing the underrepresentation of women and people of color in the industry is not just about making sure we have the necessary people. It is also a matter of social justice, and indeed creating better products and services. Tech jobs are extremely highly paid, provide excellent benefits, and often offer flexibility in terms of the number of hours worked or the ability to work from remote locations. It is inequitable that white men have disproportionate access to such jobs. Research also shows that group diversity leads to better outcomes, and diversity in leadership produces better organizational outcomes.

If we're serious about fixing the talent shortage, we should start at home.

Sextech Trends

Jessica Lipsky

High tech sex toys are nothing new, and some are even household names – *Sex and The City* brought the dual-stimulation Rabbit vibrator into public consciousness, and had a separate episode that showed a handful of women shopping at Sharper Image for "back massagers;" *50 Shades Of Grey* caused sales of kegel balls to increase several-fold. But if you talk to a sextech developer today, the vibrators in SATC and on the shelves of many stores are from the dark ages: huge, penis-shaped, resembling animals, and often marketed to men for partnered use rather than for solo masturbation and other use cases. Even at female-owned shops, sex toys can be inaccessible or intimidating for some – many of whom are looking for a more customized experience.

A growing number of startups are attempting to fill this void with high tech, DIY, and upscale devices that target people who aren't intimidated by bringing new technologies to private places. Entering into the multi-billion dollar adult toy market may pose a number of unique challenges, but several developers and executives think the time is ripe for a sexy sea change.

"I think it's people being way more honest and curious. I think there's a lot more people talking about sextech and innovation in the sex industry, which is awesome because it means people are talking more about sexuality and expression," said Jen McEwen, co-founder of adult app store MiKandi. "More people are comfortable talking about sexuality, which means technologists are comfortable making it."

Paul Clifton, co-founder of Georgia-based open source vibrator startup Comingle, believes the increasing social acceptability of "alternative" sexualities will drive a new market for high-tech devices. Add to that the coming of age of so-called digital natives and the availability of cheap development hardware, and you have a perfect storm for a creative climate.

"There's the coming of age of millennials and digital natives, people who are used to technology and use it in every single aspect of their lives. There's the acceptance of diverse sexuality plus acceptance of technology," Clifton said. "It's a brave new world, [and] they have to figure out for themselves what sexuality means to them."

The excitement around exploring sexuality online might be the result of pubescent implanting from the generation before digital natives – people in their 20s and 30s now who grew up sending sexy messages in chat rooms and continue to sext today.

"I think as we communicate more and more online and through our phones, we will continue to and have more sex totally virtually," sex and tech writer Emma McGowan said. "Right now it may be dirty texts, or dirty chats, or dirty Skype calls, but I think that will continue to evolve."

Connectivity Everywhere, Especially *Down There*

Sextech – a term that's only recently come online (pun and double entendre intended) – has broadly taken off in a few directions. The largest trend involves adding sensors and Internet connectivity to vibrators/dildos to track arousal or create a custom device, while longer-term work on teledildonics targets remote sex or remote-controlled devices. Clifton places Comingle somewhere in between; his company has developed an Ardiuno-based multi-vibrating dildo called Mod and a simpler vibe called Zildo. Comingle offers DIY kits for both, which can be purchased online or tackled in person at conferences such as Ladyfest Atlanta.

"Sexuality is so diverse and there's so many different needs and desires that probably go unmet all the time. Making open source hardware and software in a way that anyone can engage with makes it possible for a whole lot of those sexualities to be supported," Clifton said. "Hopefully people learn a little bit about the value of open source tech and how the technology they're engaging with works at a different level."

Berkeley, California-based SmartBod wants to engage techies and luddites alike with a vibrator and app that tracks arousal. The Arduino-based platform is slated for production in 2016 and combines statistical machine learning methods with temperature, moisture, and pressure sensors to track time to orgasm and when sex would feel best based on a variety of factors. Users can then program their vibrator based on their unique sexual preferences and have data to facilitate a conversation about sex.

SmartBod team working in the lab. Source: SmartBod

Other companies such as Vibease want to mentally stimulate users with a wearable vibrator that hums alongside audio erotica, but also dips into the realm of teledildonics with a remote control feature. The idea is to help keep intimacy over long distances -- many users of the wearable device are military couples where a person serving overseas controls their partner's vibrator back home, Vibease co-founder Hermione Way said.

Our increasingly connected world makes it easier to maintain sexual relationships over long distances, and might be pushing people to teledildonics.

"People who are under 30 overwhelmingly get it. This is a generation where their partner moved away to go to a different college and they wish they had this – they were on Skype every night. It's something they almost expect and wish was readily available," said Seth, a co-founder of teledildonics company FriXion. "It's a technology that I think will trickle up. I don't think it will take an entire generation to age."

FriXion has developed software to provide haptic feedback - a technological response that recreates a sense of touch, like how a video game controller vibrates when you shoot something in a game - from one sex toy to another using a 3D-force feedback joystick. FriXion aims to attach any sex toy to a virtual sensory network that feels and responds like in-person sex – contractions, vibrations, warmth, and all. The company hopes to eventually include virtual reality eyewear like the Samsung Gear VR or Microsoft HoloLens to create true augmented reality.

"We're kind of adding a higher ceiling to online flirting. It gives it somewhere to go between a video chat and actually being in the same room... beyond just masturbating at each other," Seth said, adding that users could network FriXion-enabled devices for parallel group sex. "Think of [porn star] James Deen – it's very possible with our technology that he's able to mirror his interaction across a dozen, a hundred, a thousand different people at the same time."

Seth hopes the technology behind FriXion could reach beyond the sex toy market to medical arenas, where doctors could do remote surgery using the joystick's millimeter-accuracy.

Comingle's Clifton also sees a future in teledildonics and sensory control that would translate sensor data into meaningful stimulation like vibration or warmth. "The more research that gets done, the more people realize that sex isn't just about stimulating genitals, it's about mental and emotional state, about various erogenous zones." Still, while such easy connectivity technology would be unimaginable 10 years ago, Seth said people get anxious when presented with the idea of using existing virtual networks to have sex. "Their defensiveness limits their imagination," he said.

Busting Through Barriers

Defensiveness, reluctance, and social pressure are not only issues to acceptance of teledildonics, but to smart sex toys as a whole. It's incredibly difficult for a sex toy startup to get funded by traditional venture capital firms – even in the progressive Bay Area – because investors don't want their names or firms associated with sex. The only sex toy startups that get funding outside of angel investment are those that go the medical route, MiKandi's McEwen said. MiKandi's Tits & Glass app for Google Glass was shut down weeks after its inception, and Google subsequently added a development clause that prohibited porn.

"It seems there's a big disconnect between the public who seems to be more and more supportive of sexual identity and exploration and the people with the money. Google is afraid of media and morality, so they banned all porn in Google Words," McEwen said. "Other people's sexuality is terrifying to people. It's very frustrating to experience this – we see it all the time with gay marriage."

Sextech companies have to work twice as hard to be taken seriously, she said, adding that these startups must have an "unrelenting commitment to transparency, as well as honest and great technology." Sextech startups face other traditional startup challenges, including manufacturing difficulties beyond the prototyping phase (which is increasingly done with 3D printing.) Still, many vibrators are manufactured by Chinese companies that are making it easier to plug and play.

"I think what the sex toy industry needs is a better way to manufacture and produce smaller batches. When I worked in China, it was expensive to invest in a new sex toy. The minimum order had to be 25,000 pieces. It makes it hard for a small business," McEwen said, adding that cultural barriers made manufacturing overseas difficult.

Privacy also plays an enormous role in the rollout of higher-order sextech, especially for those devices that are controlled remotely. FriXion's Seth said his company has put extra effort into encrypting and credentialing communications between networked devices, and added that safety is still a real concern.

"It's kind of challenging to think, what if someone was eavesdropping on your partner and they have a parallel experience without you knowing about it?" he said.

The reality is that emerging sextech also means new attack vectors for predators and abusers, and sextech platforms need to confront issues of privacy, anonymity, consent, sexual assault and even rape in the new sexual landscape they're creating.

Clifton notes that making communications as anonymous as possible through browsing software like Tor (which reroutes Internet traffic through thousands of relays) may be helpful in thwarting nonconsensual virtual sex, or preventing a "man in the middle" attack of connected sex toys.

Sextech: The Next Generation

Conscious of the emerging opportunities (and risks) of a new sextech landscape, the industry sits at the dawn of a new era: "I think sex toys will continue to get more and more acceptable unless we have some enormous social upheaval that turns our comfort with sexual discourse gains backward," said sexologist, author, and sex-positive activist Carol Queen, nodding to conservative politicians. "If sexuality becomes a more accepted locus of technological innovation and brainstorming, the next thing will certainly be more of teledildonics and technology where the brain-body connection is exploited."

The industry must also forge ahead in developing technology for older people, LGBTQ populations, and for people with disabilities. Mainstream company Sportsheets recently unveiled custom toys for veterans with disabilities. [See Vice article "This Company Creates Custom Sex Toys for Disabled Veterans," by Erin Meisenzahl-Peace.]

With so much technology at their fingertips, and even more sextech to come, Queen said it's of the utmost importance to give kids and teens a thorough sex education.

"Young people are growing up so technologically enabled to look up whatever they want, to get turned on to whatever they find if they're wired that way, and they're still living in a culture that refuses to give them good sexual education," she said. "If you don't know a little something, separating the wheat from the chaff in the Internet world can be detrimental."

Sextech companies are forging ahead despite cultural, funding and manufacturing hurdles. There continues to be community interest in creating and hacking sex-tech -- from sex toy hackathons to mountains of letters from users eager to get their hands on a programmable SmartBod that will help them overcome sexual challenges. If society continues on this trajectory, smart sex devices won't be just another flash in the technological pan.

You Missed a Spot

Beatrix House

American society has the best intentions when expanding tech education... or so it seems. Non-profit initiatives encouraging children, especially of color, to learn coding are numerous. Non-traditional education programs such as bootcamps have expanded outside the San Francisco and New York City tech hubs. The White House has introduced a $100 million initiative to transform an unskilled workforce into over half a million vacant tech jobs.

Best intentions, huh? Have you heard the one about the road to hell? There are still far too many potholes in the road to an IT occupation. The potholes are systemic discrimination and erasure, existing tech culture, ignorance of economic realities, and lack of quality control in education. There are still too many unanswered -- and unasked -- questions in the learn-to-code movement.

Who's Missing From The Metrics? What Is The Cost?

Last summer, Course Report released survey data about graduates from programming bootcamps. This survey, though a great start, exemplifies what metrics of representation across the industry are still missing.

The majority of space in tech careers and education is taken up by middle-class (or higher) cis white men. When someone screams "diversity," the tech world's trite buzzword, the door is left slightly ajar for cis white women.

Everyone is young, college-educated, and able-bodied with a clean rap sheet. These people become the curators of tech culture and create products to be consumed by the world. However, the world does not look like them or have the same experiences.

Since curators of tech culture largely do not represent society as a whole, the door must be ripped off its hinges to allow more people in. The tech sector needs to acknowledge that race and binary gender are not the only metric factors. Trans, queer, gender non-conforming, disabled, low-income, and middle-aged people, mothers, as well as ex-offenders and other marginalized groups consume technology. Their participation is surveilled by tech companies who appropriate ideas and innovations, then regurgitate them in the form of products, advertisements, and policies that prove dangerous to the user. (I'm looking at you, Twitter.) By doing this, companies erase entire communities, ENTIRE LABOR FORCES, who should be joining the field and changing the exclusive culture to create safer and more proficient products.

TechHire: Cure... or Tourniquet?

TechHire is a White House initiative attempting to expand coding education through traditional and non-traditional methods in a short amount of time. According to the website, its intentions [excerpted for brevity]:

• Using data and innovative hiring practices to expand openness to non-traditional hiring: Having a data-driven assessment of employer demand is critical to building a successful regional strategy.

• Expanding models for training that prepare students in months, not years: Communities will recruit, incubate and expand accelerated tech learning programs – such as coding bootcamps and innovative online training – which enable interested non-tech-experienced students to gain coding skills in months, not years.

• Activate local leadership to connect people to jobs with hiring on ramp programs: Communities will build local strategies and partnerships to connect people to jobs, with steps ranging from investing in and working with industry-trusted organizations, which will vouch for those who have the skills to do the job but who may lack the typical profile of degrees and career experience.

TechHire acknowledges a great lack of diversity in tech education and industry. To its credit, TechHire recognizes that training is a major part of this problem. Cool, but I still have questions: How can the government enforce any of this? What happens if the funds given to communities for tech education aren't used for that purpose? How is someone supposed to support themselves while participating in an intensive program? Will they receive living or food stipends? Can individuals receive government grants for this education, or will they have to compete with hundreds of other people for one scholarship?

The post-educational struggle is real for marginalized folk. How can TechHire assure that marginalized people who partake in non-traditional programs won't be written off as not "fitting the culture" when they try to get a job? This happens even when they are armed with a bachelor's degree, and the government hasn't done anything about *that*. How many of us are actually going to get a job, and not a low or non-paying apprenticeship? Is there going to be an income gap between non-traditionally trained folk and those who hold four-year degrees? Will a non-traditionally educated person ever be a contender for a leadership position? Too many people will have to try and fail in order for TechHire to find out.

And What About The Supplies?

Have you ever looked at the comments section of a tech

group on Meetup.com? I guarantee about six people will say, "I only have a Chromebook, is that okay?" This is often met with silence.

What good is expanding education in the first place if people don't have the supplies, or are maligned for such? Mac and Linux-based computers are preferred, the former being the gold standard. They are also expensive as hell. What suggestions does tech ed have for students to get what they need to succeed in their programs? Have they considered using money from well-funded Learn to Code initiatives to get laptops for their students, or working with computer companies to create a student discount or non-credit based payment plans?

Despite the popular idea that coding education is accessible to anyone, we keep ignoring the economic realities of learning to code... and that access to computers, wifi and other base supplies are real barriers to entry.

Can You Bullshit a Bootcamper?

The most popular means of non-traditional tech education is coding bootcamps. These immersive programs go from 12-18 weeks on average, and make grand promises about post-graduation possibilities. Many of them swear you don't even need coding experience to get in. For the nominal fee of $10,000, they will transform you from a coding newbie to intellectual! This is bullshit, my friends. Such an unnecessary lie, too.

If I don't need coding experience to get in, why are you giving me a technical interview? How can I Google something when I don't know what the hell I'm looking for? Mr. Bootcamp Man says I don't need to get anything right, he just wants to see if I can think critically and logically. I suppose he doesn't have the cognizance to recognize it when he speaks to me or reads my essay. I can't imagine what qualifies him to judge my answers. In fact,

I can't imagine what qualifies any admissions counselor or teacher to be on bootcamp staff. You can have all the experience in the world, but it's useless to me if you don't have teaching experience.

Don't you think we should be asking these questions, especially since coding bootcamps are now receiving government funds to expand education? It should be publicly acknowledged that bootcamps are businesses; they are for-profit schools. Yet the only regulations they are subject to are student reviews on Quora. If these businesses are going to continue to receive money from the government, they need third-party accountability.

What Happens Next?

There's still an immense amount of work that needs to be done to expand tech education. First, we need more inclusive data. It is futile to measure progress and diversity within the economic and social hegemony. We have intersecting marginalized communities who aren't being counted, and to ignore them is to erase them. Safe and comprehensive tech products cannot exist without their inclusion and leadership. We need more Angelica Rosses in tech!

Second, TechHire and other initiatives should help alleviate the financial burdens we undertake during our non-traditional tech education. This could mean taking a portion of that $100 million fund and putting it in the hands of individuals as well as programs. The money can be used for childcare, food, and housing stipends. The tuition for bootcamp and a new laptop (if needed) can cost up to $20,000. Can these educational costs that are used to further careers be tax deductible? Will companies like Capital One, who work with bootcamps to offer scholarships, also offer payment deferment to eligible recipients?

Next, we ought to have a uniform definition of what basic tech knowledge means and direct people to learning it first. Classes often require students to have basic computer proficiency. That could mean I know how to turn on my Windows XP, run a Toshiba laptop, use Chrome, and create a Word document. This is not the same as knowing basic HTML or CSS. It's important to dissipate whatever educational privileges that may exist on that first day of class. We don't send cereal chefs to throwdown with Bobby Flay, so let's not blindside beginners who are trying to learn.

§

I am an almost 30-year-old broke Black woman, five-time college dropout, and coding novice trying to learn web development. I'm using all the free online resources I can find, buying every book, and going to tech meetups. There are three unfinished boot camp applications in my inbox. This process has been disappointing because I've made a significant sacrifice to save money and learn these skills, but there still doesn't seem to be a program that is a suitable fit for me. The online resources and meetups just scratch the surface of coding languages, and don't give students insight into how a real development environment functions. The price tag on boot camps aren't attainable without a generous scholarship, and loans are not a viable option for me.

My city has a free, full-time eight-month web development program that caters to adults without four-year degrees, but I'm too old, as per the program's guidelines, and would not have had legal means of supporting myself if I joined. I recommended it to a displaced woman who has been out of work for over a decade. She said, "I can't wait eight months. I need work now."

If these tech initiatives don't work for us, who are they supposed to work for?

So You've Noticed Trans People Exist... Now What?

Social Media and the Gender Binary

Riley H.

With the increase of transgender women of color in mainstream media, social sites find themselves struggling to keep up with the "gender times." While a few major players have taken the lead by updating their gender options to include as many as 56 different gender identifications, we've only just scraped the edge of the problem, and in some cases it's clear that no transgender people had any input on the so-called improvements.

The average site hasn't evolved past "male" and "female" drop-down boxes. (Ask me how much I hate drop-down boxes for gender.) People often claim that these are the "biological" genders, focused on a false penis/vagina dichotomy, despite there being a whole scale of intersex genitalia and more than the simplistic XX and XY chromosomal formations. Nearly all dating sites force you to pick one or the other, and even OKCupid, considered more queer-friendly than other sites, only updated its options six months ago.

Even sites that have done gender updates can be shortsighted or feed into popular narratives while erasing much of the nuance that can go into a crafted gender or sexuality. For example: the emphasis on separating sex and gender, or "biology" versus "identity". Articles on the inclusion of new genders on social media sites *love* to start off defining these things as if they are different, despite the protests of many trans people who do not fit the "born in the wrong body" narrative, and leading to biological essentialism (the idea that who we are is largely a result of genetics and physical makeup) as a first introduction for many who don't know any better. It's still common to associate penises with being "male" and vaginas with being "female", despite this being a material misgendering of many trans people, and ignorant to their everyday realities.

This also leads to unnatural attention drawn to trans identities and their association to genitalia. While "trans" can stand alone to refer to the general community, it's usually an adjective that modifies a noun; for binary trans people, it describes the *state* of their woman- or manhood, the same as one would use green to describe leaves. But these terms are often merged into "transwoman" and "transman", removing woman or manhood and instead making it "transwomanhood" or "transmanhood", separating it from cisgender in a way that leaves trans people, and especially trans women of color, ostracized. It can also leave them in danger: there are people dedicated enough to create whole blogs just for outing trans women in real life.[8]

While sites like Facebook have tried to balance out this attention by including terms like "cis man" and "cis woman", trans people have expressed concerns about being forced to declare themselves transgender in an age of online violence, where trans women are constantly accused of "tricking men into thinking they're women," and where Black and Latina trans women are murdered at higher rates than any other group of people. By the end of February of this year, seven trans women had al-

---

8 "My Response To Cathy Brennan," Amy McCarthy, Bustle

ready been murdered.[9] Forceful outing is no better than complete ignorance, especially when anyone paying attention to trans women online can see the leagues of harassment they get just for existing.[10]

The "Not Normal" Category

Stepping away from the binary (because gender, like sexuality, is a spectrum!) causes an entirely new set of issues, ignorance and problems. Some trans people identify as neither man or woman, but instead any combination of such, entirely genderless, or a specific gender outside of the common ones, known overall as "non-binary". Trans people of color may use cultural genders such as hijra, koti, and two-spirit (which should only be used by the associated cultures and are not a free-for-all gender-wise!), and furthermore the usual pronouns of "he" and "she" may not apply.

Despite that, the two-genders-only model has largely stuck on smaller and newer sites. Even when given a generic "trans" option, it's often meant to refer to binary trans people, once again making "trans" a gender by itself instead of an adjective to describe a state of gender. Other sites, like Google+, provide an "other" option. It's some consideration, but it lumps together numerous other genders that exist and are just as different as man and woman, bludgeoning through cultural identities and delegating everyone who isn't a man or woman to that very dangerous "not normal" category.

Then comes the matter of pronouns. "They/them/their" are the most commonly used non-binary pronouns, and

---

9 "7 Transgender Women Have Been Murdered In 2015 And No One Is Talking About It," Katie Speller, MTV News
10 "[CENSORED BY TERFS] When the hate becomes personal: my experience with TERF harassment," Dana Taylor, TransAdvocate

even though they've been used as singular gender neutral pronouns since Shakespeare, people are still complaining about seeing them on their favorite social media sites. Further, adding only "they/them/their" pronoun options works similarly to the lone choice of "other", leaving out numerous other increasingly popular pronouns, such as the Elverson "ey/em/eir" pronouns made in 1975, and the more modern "ze/hir/hir" pronouns.[11]

Sexuality also becomes blurred when non-binary people are concerned. Terms for sexuality like "gay", "lesbian", "straight" and their formal counterparts determine sexuality relative to one's gender. With searches often focused on binary gender-only definitions of sexuality, dating sites end up oversimplifying both gender and sexuality based on... you guessed it, what you have between your legs.

I'm a genderfluid femme. My gender meanders through boy and girl and shades of grey in between somewhat freely, and I present femme as an aesthetic. I do not like to reveal my genitalia and I refer to myself as having one set just as much as another. If I prefer girls, does that make me gay or straight? What if I prefer other non-binary people, but only those of a specific type? Even on a site like OKCupid, which has done its best to allow for numerous choices, it warns that those who do not choose "man" or "woman" as at least one of their options will have trouble being matched... leaving me to misgender myself or waft alone in the non-binary void, missing out on people I may actually be compatible with.

On top of that, none of this addresses people who identify, wholly or in part, as men or women but use differing pronouns due to the nuance in their specific versions of those genders, or who identify as non-binary but do want to use "he" or "she".

---

11 Hir is pronounced like "here"

Beyond Mere Inclusion

How can sites that want to stay open to the ever-developing trans communities fix this? Just let people fill in the blanks on their gender and their pronouns, and to avoid misspelling, provide some built-in common genders and pronouns for search purposes. For searching purposes on dating sites with large amounts of queer members, a slider of gender would be lovely, like the color wheel we're all accustomed to using, which would also allow those who identify as multisexual (attracted to many genders, but not all genders), and even those who don't, to select thresholds for their matches.

Beyond mere inclusion, however, social media and dating sites simply cannot continue to ignore the increasing rates of online abuse[12] aimed at the hypervisible trans-feminine community online,[13] either. Acknowledgement is the first step, but *protection* of these vulnerable people must be the next step or it's all useless. Social media is often a place where the oppressed can discuss aspects of their lives that they cannot at home for safety reasons, a place where they go to find community. Meanwhile, social media sites are trying to become more connected to the "real", which in this case means your government identification, regardless of its relevance to one's current circumstances, leading to policies like Facebook's much-despised insistence that "real names" be used on user accounts. Google+ did the same, only abandoning the requirement after three years.

To a trans person who has not yet changed their name (or even one who has but needs to connect with transphobic

---

12 "Trans women face incessant attacks online," Sophia Banks, Canada.com

13 "The Transgender Drinking Point: Here's To the Trans Femme Revolution of 2015," Shaadi Devereaux, Black Girl Dangerous

people and thus uses their old name), this is a forcible outing. A site that can include 56 different genders but not bother to understand the nuance behind chosen names only shows a surface understanding of diversity, and it tells transgender users that their safety is not important. Though both Facebook and Google+ changed these policies in 2014, the lack of insight into the need for privacy for the transgender community is not limited to those instances. For example, Google+ completely flubbed on their synchronization when integrating with Android's KitKat OS and YouTube, causing names connected to one account to end up crossed over to other Google identities. To someone for whom protection is not an issue, this may seem like a small problem, and I have no doubt the developers at Google didn't think much about it. But many trans people can only be themselves online, where anonymity protects them from potential job loss, homelessness as well as physical and emotional abuse. Having information from accounts shared with real life family, friends, and work spliced with information revealing their trans status is, plain and simple, outing.

Despite being unintentional, when considering that very few states in the US protect trans people from discriminatory practices, there's no functional difference between this coding flub and a malicious outing by a dedicated abuser. On top of that, deleting or deactivating the related account doesn't guarantee that the information is removed from public view, leaving trans people at the discrepancy of busy customer service departments.

A Growing Iceberg

The addition of pronouns and gender words is simply not enough. If social media and dating sites are to be truly accepting of the trans community -- *all* of it and not merely the novel cases that end up in mainstream media -- then they must also accept that better privacy measures are not an optional addition.

Twitter was recently lambasted for a new feature allowing delivery of direct messages from any user -- while opt-in, many fear it will prompt an increase in the harassment Twitter itself admits it "sucks at dealing with". Tumblr doesn't even have a real block feature, allowing blocked users to see the blogs of those who've blocked them, and even reblog their posts! In fact, because the "ignore" feature only hides *notifications* from a blocked user, an abuser can hold a full public smear campaign while the victim goes unaware. Talk about making it *super easy* to harass the marginalized: forcing users to either deal with the harassment so they can keep track of its intensity, or hope it doesn't expand. On top of that, online abusers tend to travel in groups: allowing blocked users the freedom to reblog means notes from hundreds of violators who haven't been blocked yet, making Tumblr's "ignore" possibly one of the worst pseudo-blocking features ever implemented, and driving marginalized users from Tumblr.[14]

In the end, without safety measures, no amount of letting people pick the gender and sexuality terms that suit them will keep them as users. What good is having a choice of correct gender without having the privacy tools necessary to prevent the online abuse that comes with being trans, and particularly, a trans woman? Will the correct pronoun choice save the job of a trans person who has been outed through careless coding? When so many barriers exist making it difficult for trans people to change their government documents, how do pronouns help a trans person who cannot even use their chosen name? And in the case that someone is in danger, and needs to deactivate or delete an account, what good will sporting the proper pronouns on a page that may be unsafe to display do?

---

14 Personal statement, @bad_dominicana, ThisTweetCalledMyBack.tumblr.com

Allowing for proper gender selection choice and pronouns is a step in the right direction, don't get me wrong. But as it is, the steps social media sites have been making are only cracking the tip of an iceberg that transphobia has been growing for many, many years.

Q&A with STEM Professionals with Disabilities

Alice Wong

On April 14, 2015, the STEM Career Showcase for Students with Disabilities took place at UC Berkeley's Lawrence Hall of Science. The event, a joint project of the Smith-Kettlewell Eye Research Institute and the Lawrence Hall of Science, modeled itself after a STEM showcase organized by the North Carolina Museum of Natural Sciences.

The event featured a keynote speaker and a panel featuring people with disabilities currently working in various STEM fields. These panelists shared their experiences and gave attendees an understanding of the possibilities for careers in STEM for people with disabilities, in particular youth with disabilities. Attendees also had a chance to experience hands-on science demonstrations such as tactile math puzzles, and an opportunity to connect with role models in the field.[15]

Here are excerpts from interviews with three invited speakers to the STEM Career Showcase about their education, current work, and why people with disabilities are

15 For an overview of the STEM Career Showcase with interviews from students with disabilities, read "Disability No Barrier to Science" by Sarah Hillenbrand in the Berkeley Science Review (April 17, 2015). The entire presentation, "STEM Career Showcase for Students with Disabilities," can be watched at https://www.youtube.com/watch?v=_3hNKwaKMoc, published by The Lawrence Hall of Science.

an undervalued and underrepresented group in STEM. Some of their answers have been condensed for space.

Keynote speaker: Ronit Mazzoni, genetic counselor at Santa Clara Valley Medical Center and a 2005 National Federation for the Blind scholarship winner.

Panelist: Dr. Josh Miele, Associate Director of the Smith-Kettlewell Rehabilitation Engineering Research Center on Blindness and Low-Vision, Principal Investigator of the Miele Lab.

Invited speaker: Alex Ghenis, Policy and Research Specialist at the World Institute on Disability currently researching how climate change will affect people with disabilities; and founder of New Earth Disability, a blog on disability, climate change, and sustainability.

Alice Wong [AW]: What excites you about the work you are doing? What accommodations or adaptations do you use, if any?

Ronit: I love working with people and genetic counseling is a very people-oriented field as well as very science-oriented. I like helping people understand their options when it comes to their medical care and I love the feeling of being able to help people. I use a lot of different accommodations from screen readers and Braille note-taking, to labeling charts with broiled index cards. I also have a computer program that will draw a pedigree, which is a visual representation of a person's family tree which is drawn during every genetic counseling session. I also have used a hired reader to read me charts in a previous job.

Josh: I invent cool technologies that give blind and visually-impaired people better access to the information they need in order to do the things they want to do. I love it because it gives me a chance to play with really fun technology while improving my own life and the lives of my fellow blind people. The act of designing, building, experimenting, evaluating, and otherwise innovating is also thrilling to me. It's like solving puzzles that have real impact in the world -- my world. I use several different screen readers, Braille displays, phones and tablets with adaptive software, and a cane.

Alex: I love what I'm doing because it combines my passions for climate change issues and disability rights... It's actually interesting, right out of college I was working in the energy storage field doing renewable energy policy - which was more of a "STEM" career... It was super exciting and I loved it because it was creating a greener grid that would be more resilient in the future. Now I'm working on those climate change and social justice issues more directly and I'm much happier with my job.

AW: Describe any challenges you faced during your education, training, and search for employment.

Ronit: My first challenge was convincing the graduate programs during my interviews that I would be capable of being a genetic counselor, even if I didn't have all the answers as to how I would accomplish things at that moment. My next challenge was figuring out how to draw a pedigree independently. My biggest challenge by far has been convincing others that I am capable. I received a bad grade on one of my rotations due to close-mindedness about my abilities and it took me over a year to find my first job and over six months to find my second.

Josh: Simply too many to list... None of my books were ever available, getting notes was difficult and time consuming, managing readers was a job by itself, getting access to charts, tables, graphs, etc. was almost impossible, as was creating them.

AW: What are some common misconceptions and stereotypes regarding STEM education and careers for people with disabilities?

Ronit: People in general do not think people with disabilities are as capable of doing anything, this includes simple life skills like dressing, cleaning the house, picking out clothes, taking care of children, etc. If a person with a disability cannot do these things, how could they possibly do math or science? I believe people who are not disabled imagine themselves if they were disabled and could not imagine how they would accomplish these tasks, so they automatically think it's not possible.

Josh: A lot of teachers think it's going to be difficult to create an inclusive curriculum or to otherwise provide an accessible educational experience for students with disabilities. This is especially true in lab classes like bio or chemistry. The truth is that this attitude is more of a barrier than anything else. There are usually relatively simple ways of adapting activities and assignments to allow students with disabilities to participate.

Alex: The main thing is that people think that people with disabilities have trouble doing any jobs, STEM or not... People also view us as just not being sharp enough to do things that require hard skills... stereotypes keep us from pursuing STEM education and careers, they make it so that college admissions and hiring managers don't let as many of us in, and then people in general don't see us in those types of careers so the stereotypes get reinforced.

AW: One of the biggest barriers (in my opinion) is the sense of low expectations for students with disabilities — how can educators, parents, and society in general encourage students with disabilities who want to pursue their interests in STEM?

Ronit: I think it's most important for students with disabilities to find others who have disabilities who have done it. That is the biggest and most effective way to prove that it's possible. If you can't find someone who has done it, think outside the box and be creative in solving problems.

Josh: I'm not as worried about the students who are driven to succeed in STEM -- they are likely to find a way. It's the kids who don't really care that much about school or who want to do the minimum of work to get by who are the most at risk. Non-disabled kids who do not have a passion for learning are often able to get enough out of their education to be able to succeed in a STEM career once the apathy of childhood wears off. Children with disabilities in this same situation are often allowed by similarly apathetic teachers and overwhelmed parents to isolate themselves from peers and learning such that they are at a severe disadvantage later. By the time the kids realize that they might be interested in a technical field it's too late. Low expectations are at the core of it!

AW: What were your experiences like as a young student (before college) learning and experimenting with science? What barriers, if any, did you experience and how did you deal with them?

Ronit: I used a helper in all my science labs and learned to ask questions and get descriptions of what was happening. I asked for help a lot when things were visual and I didn't understand concepts.

Josh: I had a lot of social problems as a blind and burned little kid. Many other kids treated me very badly and before high school I was constantly being bullied and getting into fist fights. At the same time, I was really lucky and had excellent resources, teachers, friends, and parents. Nobody that I cared about ever told me that there was *anything* I couldn't do. I had a teacher who transcribed all of my assignments into Braille and drew all of my tactile graphics for me. I had teachers who realized I was smart and had high expectations of me. I had parents and relatives who were educated and could help me find answers to my questions and were devoted to my success. I had it so easy, and still it was pretty hard...

AW: With the numerous diversity efforts in STEM, what unique skills and perspectives do people with disabilities bring to these fields?

Ronit: I think creativity and perseverance are two things people with disabilities can bring to STEM fields. That is what I feel I can contribute the most to genetic counselors and patients I see.

Josh: I think it really depends on the specific field. In my case I help develop the technology used by other blind people. A sighted person simply would not have the same insight and understanding about the need, use cases, attitudes, and applications that I do as a blind researcher. I'm not sure if a disabled perspective has any impact on the pure sciences like theoretical physics, but it definitely can on applied sciences where people with disabilities will be the beneficiaries or recipients (or who will have to live with the barriers) of the tools and techniques being developed by STEM professionals.

Alex: STEM topics impact us more personally than able-bodied folks: most of us rely on medical equipment that relies on solid engineering and construction to work, not to mention functioning support systems like a solid electrical grid. So we can bring a perspective of how important STEM is to people out in the real world, and that carries extra passion along with it. As for skills, a lot of us spend time working with that same equipment, so we know the intricacies of engineering better than a lot of people... Insight is a valuable skill.

AW: How can STEM industry leaders, employers, and educators better recruit and accommodate people with disabilities into the workforce?

Ronit: I think being open-minded is the only barrier to people with disabilities being employed or doing well in school. I have only done well when people around me are open-minded and most of that is due to my perseverance and convincing them as best I can to give me a chance to prove my skills. This, however, is very difficult and can be very depressing. If people got more education about people with disabilities, perhaps disabled people wouldn't have to work so hard to convince others.

Josh: Seek out qualified applicants with disabilities, hire them, and make sure they have the accommodations they need to contribute. We're out here -- just let us in and let us work!

Why We Need People with Disabilities in STEM

Many people assume that it's not 'safe' or 'accessible' for students with a variety of disabilities to pursue STEM for study or career (or for pure enjoyment). With such a need to increase student engagement in science and the STEM workforce pipeline, students with disabilities can play a role in filling this need. Yet in the current discussion on diversity and STEM, as with so many diversity initiatives and programs in various fields, disability is usually excluded or thought of purely in terms of accessibility or accommodations.

Reimagining disability as diversity, here are some important points made by Ronit, Josh, and Alex on STEM and people with disabilities:

• Social attitudes and discrimination are barriers just as disabling as physical inaccessibility or lack of services.

• Low expectations of students with disabilities can have a detrimental effect, narrowing future life chances and options even further.

• Adapting activities, curricula, and assignments is not as difficult as it seems; it merely requires imagination, creativity and flexibility.

• People with disabilities have always been problem-solvers and adaptable in difficult situations; the outsider perspective can bring valuable insight to STEM fields.

• People with disabilities have had to prove themselves over and over again to their non-disabled counterparts that they can do the same work just for the opportunity to be considered or included.

• People with disabilities can provide unique viewpoints in all STEM fields, from their user experience to their skills and ability to 'think outside the box.'

These barriers are very real and can profoundly shape the career paths of people with disabilities who otherwise would be involved in STEM. Heather De Mian (@ MissJupiter1957 on Twitter), an independent videographer who uses a wheelchair, tweeted about her early interest in paleontology and forensic anthropology in a conversation on women and STEM with @AJStream on March 17, 2015:

Was going 2 B paleontologist, but Geo Dept decided 4 me I couldn't attend required camp b/c of disability. #Disabled #WomenInSTEM

Then minored in forensic anthropology, but couldn't major b/c required lab was not w/c accessible.

Viols of fed civil rights laws. Basically had "no cripples allowed" signs on science depts.

I should've been allowed 2 pursue career w/ dinosaurs or forensic facial reconstruction regardless of how I was born.

Despite federally-funded state unis like mine being required to make their programs accessible since 1977.

[Excerpted from the full thread, available at https://twitter.com/MissJupiter1957/status/577919511304630272]

Similar to Heather's experiences from the 1990s, when I was in junior high and high school, I had to sit separately in my chemistry and biology classes because my wheelchair couldn't fit under those high lab tables. This separation kept me from having hands-on experiences with experiments and other lab activities (not to mention the social distance from my classmates). I was relegated to the role of recording and observing while my lab partner had options to participate in all aspects of the coursework.

No student with a disability should be sidelined from science education or discouraged from pursuing a career in STEM — hopefully events like the STEM showcase, mentoring programs, internships, and a growing critical mass of people with disabilities in STEM can reverse this negative trend.

Additional Resources

AccessSTEM: The Alliance for Students with Disabilities in Science, Technology, Engineering, and Mathematics http://www.washington.edu/doit/programs/accessstem/overview/about-accessstem-project

Entry Point: AAAS Project on Science, Technology, and Disability http://ehrweb01.aaas.org/entrypoint/

National Science Foundation: The Recruitment and Hiring of Individuals with Disabilities and Targeted Disabilities, FY 2011-2016 http://www.nsf.gov/od/odi/plans/NSFDisabilityHiringPlan2011thru2016.pdf

Special thanks to the interviewees Ronit Mazzoni, Josh Miele, and Alex Ghenis for their participation and to Sherry Hsi, Research Director of the Lawrence Hall of Science and Katia Albanese of ePolicyWorks for their assistance.

Interview with Amy Wibowo, Creator of BubbleSort Zines

We sat down to speak with Amy Wibowo about BubbleSort, teaching computer science, making independent tech media and building successful crowdfunding campaigns... plus taking your passion full-time!

MVC: Tell us a bit about BubbleSort zines!

BubbleSort explains computer science concepts using drawings, diagrams, cartoons, and stories. Their intent is to make computer science more accessible for people who have been made to feel that computer science isn't for them. They'll cover topics including cryptography, adder circuits, embedded programming, and caching. The covers and diagrams are all hand-drawn, and they'll be locally printed on really nice paper.

MVC: Your zines represent a fresh and creative way of learning about computer science. What limitations are there in how CS is currently taught, and what possibilities exist for new kinds of pedagogy?

I was once talking to some young women taking high school computer science and asked them about what they were learning in class. "Java," one of them said, "We're learning Java." She looked around at the others and they all nodded. I was really sad that the class had given them the impression that the language and specific syntaxes of the language were the main point of the course.

As research for BubbleSort, I examined the current AP Computer Science curriculum. It starts off with an introduction to object oriented programming. It sounded like the course curriculum was completely uninterested in appealing to anyone unless they already knew they wanted to study computer science! That's sad because high school students shouldn't have to already know what they want to study -- high school classes should encourage people to explore lots of different subjects to figure it out.

In addition, classes are losing out by focusing on pure computer science concepts, without tying them to other subjects. Finland, which is consistently rated as having some of the best education in the world, has recently decided to completely do away with subjects. Imagine classes that did things like tie in literature and computer science -- teaching students to write programs that generate sonnets while learning the rules for sonnet composition! There are so many possibilities when you connect subjects to each other.

Not all classrooms can afford one computer per student, but lots of concepts in computer science don't even necessarily need a computer. I'd love to see more low-tech courses that focus on general computer science concepts and can be taught in all classrooms, vs. courses that focus on very specific technologies or languages and require expensive equipment.

MVC: What is the zine-making process like? What are the most difficult parts, and what advice would you have for someone who is just getting started making zines?

The zine-making process, for me anyway, is really free-form. I usually start off by designing a cover, which helps guide me to write the rest of the zine. Then, I write outlines and go back and fill them in. I'm a very visual and tactile thinker -- I write best if it's on actual paper with nice pens or markers, and drawing illustrations helps me come up with what I want to say. Lots of times ideas for content come to me randomly while doing something else, and often not necessarily in the sequential order I'd expected or hoped for. Probably the most stressful parts are making myself write out of order like this -- writing Chapter 3 when I don't have super clear ideas for Chapters 1 and 2, and trusting those parts will come to me later. Or making myself write when I don't know 100% how I want to phrase something yet, and trusting that it will sound better when I go back to edit.

I picked zines as my medium because, among other reasons, they have anti-establishment roots. As the writer of a zine, you have complete control over your timeline, your content, your design. And also, no one is expecting perfection from a zine. I put off doing this project for years thinking, "I'll do this when I'm a better writer" and "I'll do this when I've gotten better at drawing." But then I realized, it's totally OK to not draw perfectly in a zine, or to write perfectly. So the advice I have for people just getting started making zines is: you are already good enough, and what you have to say is valuable.

MVC: What inspired you to pursue this full-time after working as a software engineer for the past few years? What has the process been like, and what are the major obstacles you've found?

I've been slowly building up to it. On weeknights and weekends, I've been pursuing creative side projects: knitting sweaters on a hacked knitting machine, watercoloring, writing and drawing zines. I was having dinner with a friend, Linda Liukas (the author of children's book "Hello Ruby"), who was encouraging me in my personal projects and who reminded me that I have so many dreams of my own that I'd like to pursue, but I continuously put off to work on other people's dreams. Working at startups has been a comfortable and predictable lifestyle for me, and I've learned so much from people that I've worked with, but it was definitely time for me to do something different.

For so long, I felt like I'd been fighting the system from

within, trying to make the companies that I've worked at have a better and more inclusive culture, but I definitely burned out on that. It's nice being able to pick who I get to work with and only pick people who I know I won't need to have conversations about intersectional feminism 101 with.

Money has been the major obstacle so far -- startup salaries are really comfortable and I'm definitely making less money this way, so I'm supplementing that with contract work.

MVC: One of the things you talk about often is the intersection between tech and art.

A lot of people have trouble believing that those intersections exist and can't grasp the idea that anyone could be interested in both tech and art. If people have seen things I've drawn, and hear that I work at a startup, they assume I'm a designer and are shocked to hear I'm a programmer, or assume that I must be junior. Or people I've worked with have been surprised to see my art. Lots of programmers seem to think that computer science is harder and more valuable to society than the humanities, which is just completely utterly arrogant and wrong.

Art and technology have always been really connected in my mind. I was equally interested in art and math and science growing up. I decided to study computer science because I loved making things and computer science was a way to make things, using math and science. Good visualizations can make complex math and science concepts much easier to understand. And you can use computer science to generate really beautiful art. Most of my motivations for pushing myself as a programming/hardware hacker are around making tools that enable people to be more creative: software that helps you generate clothing patterns, hacking a knitting machine to print images, a web app that gives you instructions for a Lego mural of the pictures you send to it.

Because I see computer science mainly as a tool for social impact and creativity, I feel like there are lots of missed opportunities to encourage people to pursue their interests at these intersections when we frame computer science classes around purely algorithms or object oriented programming or syntax of a specific language. We also lose the opportunity to engage marginalized people, who, studies show, are often more interested in careers based on how much social impact their work will have.

MVC: You've recently led an extremely successful crowdfunding campaign on Kickstarter. Crowdfunding is a major way that new projects in tech and diversity are being funded. What did you learn from this process? What worked, what didn't, and what would you do different next time?

Crowdfunding is really great for new independent projects since it can help you not have to take funding from VCs. It helps you assess what interest already exists for your product, and if you're making a physical good, it's super helpful as pre-orders so you don't have to invest all of your own capital. Running a crowdfunding campaign by myself, I learned the difficulty of taking on roles that, working as a programmer at a mid-to-large startup, I hadn't had to take on before. I had to be my own marketing, PR, design, photography, and finance department.

I was really excited when The New Yorker reached out to interview me about the Kickstarter -- but according to the metrics, it only lead to 15 additional backers! I was surprised how much press didn't seem to matter for the campaign so much as word of mouth.

A couple people thought I was charging too much and suggested bringing my prices down. I had misgivings about this, because I think lots of marginalized people are made to undervalue their own time and work and I didn't want to do that. The zines are going to be printed as a small run, and I am dedicating many hours to writing them. I do want to make the zines accessible to everyone who'd like to learn CS, but I don't feel like I should be bearing 100% of that financially -- I already left my software engineering job to do this. People and companies have been generous about backing at levels that donate zines to schools in place of, or in addition to, zines for themselves.

MVC: Tell us what's in the works for your next BubbleSort zine!!

I'm hoping for the very first BubbleSort zine to come out in June, and I'm really excited about it! It's going to be about how calculators work. The very first chapter is going to look at the history of numbers and counting around the world. I want the chapter to convey that computation is completely a product of history, society, and culture. Also, that there are so many different ways to represent a concept, and that there is power in those different representations. Then the next chapters will cover binary numbers, formal logic, logic circuits, and finally building your own simple calculator! I hope that this zine will help people think about computation and numbers in ways they might not have thought of before.

Check out the BubbleSort shop at shop.bubblesort.io

Does Not Compute: Destroying the Autistic Savant Trope in Programming

Catherine [Kiran/Rin] Oliver

I tried to code an app. 250 lines of code in, I realized that I had coded it in the wrong language for what I wanted to do. Not only that, but it was riddled with errors. I had been learning to code through Free Code Camp and Codecademy for two months on a nightly basis, but I couldn't hack it.

Does this make me a failed programmer? That depends on who you ask. If I tried to re-code the application in JavaScript, could I fix it? Maybe.

Individuals that are on the autism spectrum, have dyscalculia, or a nonverbal learning disorder often find mathematical processes and the way of thinking required to program efficiently (and well) close to impossible. I am 30 years old, and I would be lying to you if I told you I was able to do math beyond rudimentary fractions, limited algebra, and basic geometry. At best, I perhaps can complete 6th grade level math. Given the new Common

Core, what I consider 'sixth grade' math might be elementary school mathematics now.

This does not make me unintelligent. I have a novel being traditionally published, due out later this year and totaling over 60,000 words. I've worked in public relations, marketing, and community management for four years, in addition to interning at amazing companies such as Learning Games Network doing community outreach, PR for the Green Party (beating out thousands of other applicants) during the 2012 presidential run, and more.

The thing is, that in today's world — all that means less and less. Apps, web design, product development... today's business landscape increasingly demands at least a rudimentary knowledge of coding/programming. For writers, artists, musicians, community managers, organizers, and many others that simply are not 'wired' for mathematics or abstract thinking such as the type used in programming, many technical skills are not only difficult, but frustrating: the tools and pathways for learning these skills assume everyone is a neurotypical, able-bodied person with a homogenous skillset, capable of reproducing complex computer outputs without error.

The ableism inherent in the "learn to code" movement keeps huge swathes of the population from accessing these skills. While using Codecademy and Free Code Camp, I found these platforms assumed three things that were very, very wrong about me:

1. I had a basic understanding of 'if x = this then y = this' sort of statements. These are used in the Ruby on Rails section in Codecademy, and the Free Code Camp JQuery modules.

2. I was able to discern patterns that neurotypical people deemed 'logical' or 'next in the sequence.'

3. That I was neurotypical in the first place.

Coding with ASD and dyscalculia is more frustrating than it has any right to be, at least for me (I don't speak for everyone, and everyone's experience is different.) There are no programs, minus those for children, which explain the 'basics' of coding. Even less that 'explain' in terms which don't belittle or talk down to people: there's a subtle way Learn to Code programs speak which tells people that they just haven't looked at it 'the right way' if they don't understand how a module works.

The problem is, I don't see the 'way' to look at it in the first place, because I am not neurotypical or able to understand the pattern or solution presented.

The Slippery Slope of Being 'Beneficial to Society'

Across the educational system, there is a push to learn STEM at the expense of cutting creative writing, music, art, and design classes. Where is the pedagogy, platforms and tooling that can help people like me succeed in these subjects? 9/10 times if you ask me to do some sort of math problem, I'll get the answer wrong. I have had early intervention, IEPs that were 2" thick, calculators on tests (I still failed those) and every other 'invention' that neurotypical people thought could get me to just 'understand' math. If we had had programming classes in my school, I would have failed them. I just barely passed the standardized testing we had to do to graduate high school. I was in the 98th percentile in language and writing.... And the 2nd from the bottom in mathematics on the PSATs.

What is the future for people like me when the privileging of STEM skills, currently accessible almost solely to the neurotypical and able-bodied, is reflected not only in our educational system but across businesses, corporations, coding camps, and so much more? It leaves hundreds of thousands of people feeling as though they are less valuable or even worthless because they have not 'unlocked' the secret to coding or programming apps. The combination of cutting off access to STEM skills to neurodiverse people, and privileging STEM over all other fields, has serious consequences. When we say that only STEM benefits society, we carve a swath through those that make their living enriching the lives of others through words, music, entertainment... things considered 'hobbies' in the tech sphere. In that world, it's fine to write, or draw, or play music — as long as you grind away coding for 10-16 hours a day in a STEM field and eat lunch at your desk.

Encouraging programmers to work 16-hour days is not only unhealthy, but places unrealistic expectations on everyone in society. We are not machines. It is not a mark of "intelligence" "drive" "moxie" or "grit" to sell your soul to a job. The work/life balance for coders and non-STEM workers is often equally awful, with many millennials forced to work while at dinner, away from home, while out with their families... you name it. People in professions other than STEM are held to STEM standards in terms of our work days, what we should do in our off-time -- but coders often get cushy offices, can work from home, and have a more than manageable salary. For neurodiverse and marginalized non-STEM workers, the difference in pay and quality of life is often tens of thousands of dollars or more, a significant gap in financial security. Meanwhile, we are told that we are being lazy if we cannot complete the same work day a coder can, despite the gap in pay, privilege and working conditions.

Within this system, we tell those who cannot code due to learning disabilities, autism and having different skills sets that their time is not valued or appreciated, all while we hold them to a cisgender, white-male-dictated version of "acceptable" levels of production, quality, and output. Understanding that those who are on the autism spectrum, living with dyscalculia, NLD, or another learning disorder are not lazy, incapable, or deserving of less pay for the same amount of hours worked is crucial. We must understand that those working in non-STEM specialties are just as valuable as their STEM co-workers. The most successful app will fail if users cannot connect to it in a way which works for them. No matter how simple the features, if there is not a layer of communication, an application to how someone could use an app in their daily life and how it enriches their involvement with their community -- it will not succeed.

What Does Destroying the Autistic Savant Trope Entail?

Many autistic individuals that disclose our disability are assumed to be brilliant programmers. Often, people assume we are robotic, even computer-like in our emotions, which couldn't be further from the truth. If you know one autistic person, you know *just one* autistic person. Everyone is different, and everyone's experience with programming and computers is also different. We as a society can combat the troubling autistic savant trope and rampant ableism in programming by refusing to give in to the 'autistic savant' trope.

We also need to create tools which teach coding in a way that is not ableist. There must be, at a minimum, a way to address programming which is not so childish as to insult adult learners. To be told to use a child's application to learn to program is insulting and ableist, as we wouldn't need to be told this if standard programming tools were created with neurodiversity in mind.

One approach is to structure coding and programming in different ways. Tools could be written that demonstrate programming concepts in, for example, writing. Parts of speech, adverbs, nouns, grammar... all could be used to help linguistically and artistically-inclined autistic individuals better understand programming. I know where to put commas, I don't know where to put <print>

commands. I can blend colors, understand a color wheel. Read music. Make a program that shows me how to code in a way I understand. Not the way you *think* I should understand. Not by assuming autistic people "*should*" understand what comes naturally to neurotypical people.

How we continue to improve tech culture is up to all of us. But it starts by realizing that the 'low barriers' to learning to code are hurdles that some may never overcome if we won't confront ableism in their design.

Building Better Tech Events: An Interview with Claudina Sarahe, Co-Founder of SassConf

We got together with Claudina, who co-founded and organizes SassConf - a week-long conference for front-end developers and designers that love Sass and are passionate about building a better, more beautiful web, for users and creators. We discussed open-sourcing the event organization process, developing more inclusive standards and policies, and how to get started running events in the tech community.

MVC: Tell us a bit about the world of Sass and your role in it! What is SassConf, how did it get started?

The Sass community is truly, for me, one of the most openly queer and trans communities in tech which is not something that you normally see in the field. After the first CampSass in April 2014 – a single day Sass mini-conference run by Hampton and Michael Catlin – I was talking to Chris Eppstein, one of the core developers of the language, and he told me: we (Chris, Natalie Weizenbaum, and Hampton) designed Sass with community in mind from the beginning. It really shows. Community was never an afterthought for Sass, it was and is as important as the tool itself. The community guidelines highlight this ethos.

The idea for a conference for Sass enthusiasts has its roots in Austin, Texas. In 2012, Scott Kellum and I met up to talk about Sass at a restaurant. Scott showed me what was then a very early version of his amazing grid system, Singularity. Both of us live in New York, and at the time we were surprised we had only just met. We wondered, "what if other people want to meet up and talk about this thing, maybe we can have a conference to facilitate that." We ended up starting a meetup in New York with Sam Richard and support from Mason Wendell. Just as much as SassConf is focused on the community, its genesis would not have happened without support from community-focused local organizations such as New Work City and Pivotal Labs.

It took us some months of running the meetup plus a convincing number of email signups and prodding tweets before we finally committed and said, hey, we're going to go for this, let's bring together the Sass community.

That was 2012 and the first conference happened in 2013. I worked closely together with a team comprised of Aisha Green, Adekunle Oduye, Robert James, and Sam for the first conference and again in 2014, along with tremendous support from people in New York, including Scott, Lissa Aguilar, Jen Gergen, and a rad group of volunteers. As far as my role, I deal with the conference which is another way of saying I do a bit of everything and anything. Organizing an event is a big task, but it isn't an impossible one; I want to help those with ideas feel like it's something they can do. I'm increasingly more and more interested in designing and shaping a process to organize events and conferences that makes them more open and accessible for anyone to do. Organizing is incredibly important for the social change that we want to see in our society.

MVC: SassConf is doing some really interesting things around support, inclusion and diversity of speakers -- for example, you offer mentors for the CFP process. Can you talk about how that works?

The CFP mentor process grew out of our office hours experiment that we offered last year. It's being led by Elyse Holladay, with tremendous support from Scott. The main idea is to give people a chance to chat with someone about their talk, to suss out the questions, to figure out "hey, what can I do to make this better?" or "do you think this is a good idea for a talk, do I have enough content?" The truth is that conference organizers are curating, we're trying to tell a story and I think people shy away from submitting talks because we're worried about whether or not it'll be good, or if it's what the organizers are looking for. That means that a lot of really good ideas don't make it onto the stage and into the universe.

Our goal is to provide that extra support to help people iterate through their ideas, and have mentorship that comes from people within the larger community of front-end development, not just from the Sass community. Many are organizers that have insight into how to write a better proposal. We leave it pretty open-ended for connecting to mentors. You can book a time with a CFP mentor. You can also pop into the IRC chat room. Additionally, many of our mentors have made themselves available via their personal Twitter accounts.

I'm still blown away that every mentor has said "yes." The feedback from mentors and participants has been really positive. Elyse and the mentor team are doing an incredible job. We're all looking forward to the results.

MVC: This year, you've changed your CFP process from a blind review process to an open, transparent process.

What's the rationale behind that, and what do you hope will change?

Plainly, the blind review process didn't work for us. The second year of SassConf, we scrubbed all the submissions for names, references, pronouns, everything but the title of the talk and the content. Sam and I curated the first round. We invited about 10-15 community members as reviewers to vote on the remaining talks and we tallied results. When we revealed our final list it barely included any non-white cis male applicants.

I was shocked; the blind review process is really held up as the way to do things if you want a more diverse event. So I went back through the original list of submissions and realized that there was no way a blind review would ever be fair, because we didn't have an equal number of submissions from underrepresented and diverse candidates. There was just no way that it would work out. At the end of the day, if your applicant pool isn't diverse, then you're not guaranteed a diverse line-up with a blind review process.

That led us to give the open CFP process a shot. We were really inspired by the work of Karolina Szczur from CSSConf Oakland as well as the CascadiaJS Team, and we took a lot of inspiration from Dan Denney of Front-End Design Conference. Now we host all talk submissions on GitHub. Similar to being able to talk to someone about your ideas, it's nice to get a glimpse of what people are submitting so you don't submit yet another style guide talk or CSS maintenance talk; or if you do, you can figure out what other people aren't talking about and craft a better proposal. The ability to go back and edit your proposal in response to feedback is also really helpful. Being able to get feedback and see people discuss concepts, or finding something that you like and being able to say "hey, I would really like to see that talk"... all of those things are incredibly valuable and attainable with an open process.

MVC: Another thing you're doing is really integrating tech tools into how you organize the event. You're having people submit talks on GitHub, using IRC for mentorship sessions...

Open tools mean a much more visible and transparent organization, and at the end of the day that's what we want. We've also learned so much from other conferences and people that share their process. If we, as individuals and organizations, truly want to get better, we have to be open to feedback, and that requires being as public and transparent as possible. Additionally, I think it is valuable for us to use tools that we are already familiar with in new and different ways, exposing their true collaborative nature.

As an org, we still do a lot of our planning in a private GitHub repo -- that's just because of the sensitivity of the information. However, one of our main goals for this year, especially as the conference is being adopted by other cities, is to open source as much of our process as possible. So we're looking at putting up an example repo with docs, with our checklists, and the things that we've done to run SassConf over the last few years.

MVC: One of the things we discussed when we were setting up this interview is the SassConf mentality of trying new things, experimenting, seeing how things go and making changes based on the results. How does that philosophy pervade the way you operate?

Maybe it's my childhood of growing up in science and really loving the experimental process, but I believe that experimentation is the process that builds creativity. I definitely don't view anything that we do as set in stone or dogmatic. Every year you try and you tweak and you adjust.

The conference itself started out as one giant experiment starting with the meet-up. Our first year was an experiment in *can we do this* and bring everyone together; *are people even interested* in coming together. I'm forever thankful to every single one of the speakers that flew out that year on their own expense, prepared a talk and gave it because they wanted to be there. We would not be here were it not for them. The second year, our experimental goal was to make sure we could pay our speakers and dedicate a full day to the community unconference [or participant-driven meeting], which we did. And our main goals for this year are to represent and celebrate the diversity within our community and become more family-inclusive.

As far as evaluating our experiments, I find retrospectives, talking about how things are going and checking in, to be immensely valuable, and the organizing team aims to provide a space where that's as comfortable as possible. Things invariably don't go as planned, and sometimes it's hard to receive feedback. It can also be hard to give. So we work on creating comfortable, safe environments where we feel like we can give feedback and say what's on our minds responsibly and without fear of judgment. And we're getting better because of it.

MVC: Something we think is really cool is how diverse the SassConf team itself is! Can you tell us a little about the team, and how it's grown?

Wow. What an incredible team. I'll always have a deep fondness for the founding team in New York, which is made up of Aisha Green, Robert James and Adekunle Oduye. What can I say about these people? There's just immense love, and it's amazing to think about how long we've been organizing together and the things we've been able to accomplish. We're now working closely on GothamSass as Aisha, Robert, and Adekunle, along with Tony Bellamy and Nicholas Frota, prepare to take over leadership.

Also from the core team is Scott Kellum, who is responsible for SassConf's gorgeous brand and is also heavily involved in mentorship this year. Sam Richard organized in 2013 and 2014 but is no longer a core organizer. Last year we had immense help from Michelle Martinez.

This year, we've expanded the team as we start to think about how the conference can move from city to city. We've actually added new core positions. Elyse Holladay is chairing the speaker selection process and implementing a lot of the new ideas we've been talking about. During the SassConf 2014 speaker dinner, she said to me, "You know, I'm a little disappointed by the under-representation within our community especially given how diverse we are, and I think we can do better." And I said yeah, you're absolutely right we can do better, do you want to help make this better? She said yes. Moments you'll never forget.

Elyse is a local organizer of ATX Sass along with Ian Carrico and Una Kravets, who are also helping out on logistics, such as food — delicious food is very important — and things to do in Austin that not only highlight the city but also provide the right environment for happenstance, for the magical moments that come from gathering and collaborating.

I believe teams are really valuable and important, and SassConf is best represented by the inclusion of a diversity of organizers, thoughts, and ideas. This is how we keep it inclusive, welcoming, and uniquely fun each year.

MVC: What's been the most surprising or challenging thing you've learned doing SassConf? What advice would you have for other event organizers?

Everything is a lesson! One of the most surprising things was just the result of the blind review process and how it failed. Everything I read touted how this is the right way to do things. Just shows that there isn't really a right way, it's about figuring out what's going to work best for your needs, your community.

In terms of advice for other event organizers: set goals and establish accountability amongst the organizers. Prioritize developing and sustaining good communications channels. Value honesty and openness. Make sure to pick people that you trust and are comfortable organizing with, because it does become like a little family where you have to be able to be honest and say "hey, I'm super busy" or "I'm not feeling so hot, I need some time off," and be able to trust that people are there to keep moving things forward and commit to what they've agreed to do.

And to budding organizers, those of you with ideas, just do it! Put it in motion. If you have an idea for an event, try it out. Dream big: put out all the things that you want to do, and then pare them down. But still have that list, your reach, your absolute best best, and then your second best and your compromise, and make sure you don't go lower than your compromise.

Don't worry if it only ends up being one event. Any event is better than zero events, especially if there's good intent behind it. Besides, in the day of so many conferences, perhaps designing for obsolescence, for the ephemeral is where we should put our focus.

MVC: And finally, how can people get involved in SassConf?

Join a local meetup. Doesn't have to be purely Sass, the important part is participating in your local community.

Submit to speak at SassConf or about Sass at other conferences. Attend. Sponsor. We'd love to see you in Austin.

Contributors

Shanley Kane

Shanley Kane is founder and CEO of Model View Culture. She is a cultural critic, organizer, writer and feminist. Her writing on startups, management and culture can be purchased online in the collection "Your Startup is Broken: Inside the Toxic Heart of Tech Culture."

Nicole Forsgren

Nicole Forsgren, PhD, is the Director of Organizational Performance & Analytics at Chef, a configuration management and IT automation leader. She is an academic partner at the Social Analytics Institute at Clemson University and received her PhD in Management Information Systems and her Masters in Accounting from the University of Arizona. Prior to her role at Chef, Nicole was an Assistant Professor at Utah State University and Pepperdine University. She is an expert in IT adoption and use, DevOps impacts, and communication and knowledge management practices, particularly among technical professionals. Her background spans analytics, enterprise storage (specializing in RAID performance), cost allocation, and user experience. She is a featured speaker at industry and academic events and is involved in women in technology initiatives.

Jez Humble

Jez Humble is a vice president at Chef, a lecturer at UC Berkeley, and co-author of the Jolt Award winning *Continuous Delivery*, published in Martin Fowler's Signature Series (Addison Wesley, 2010), and *Lean Enterprise*, in Eric Ries' Lean series (O'Reilly, 2014). He has worked as a software developer, product manager, executive, consultant and trainer across a wide variety of domains and technologies. His focus is on helping organisations deliver valuable, high-quality software frequently and reliably through implementing effective engineering practices.

Jessica Lipsky

Jessica Lipsky is a reporter/editor covering hardware and consumer technology. When not focusing on tech, she writes about government, culture, and music for several pubs. You can occasionally catch her spinning soul records or cutting a rug in the Bay Area.

Beatrix House

Beatrix House is a Harlemite and middle-aged autodidact coder.

Riley H

Non-binary trans, femme homo who returned to school after many million years to devour the tech world with their zany calls for diversity, ready to smash the (imaginary) meritocracy. Anime geek and eternal Japanese language student who works on coding games, pixel art and fantasy fiction novel writing in their free time.

Alice Wong

Alice Wong is the Founder/Project Coordinator of the Disability Visibility Project, a community partnership with StoryCorps collecting oral histories of the disability community in celebration of the 25th anniversary of the Americans with Disabilities Act. Alice is also an Advisory Board member of APIDC (Asian Pacific Islanders with Disabilities of California) and a Presidential appointee to the National Council on Disability, an independent federal agency on disability policy. Among the issues she cares about are web accessibility, improving STEM education for students with disabilities, and increasing the number of people with disabilities in STEM careers. She's also a proud disabled Nerd of Color.

Amy Wibowo

Amy Wibowo is the founder of Bubblesort Zines, zines that aim to make computer science concepts more accessible through drawings and diagrams. Previously, she did web development at Airbnb, machine learning research on the ASIMO team at Honda Research Institute in Japan, and HCI research at the University of Tokyo. She also enjoys hardware hacking and the intersection of art and technology-- in particular, making tools that enable people to be more creative.

Catherine [Kiran/Rin] Oliver

Catherine/Rin is a white, trans [nonbinary], grey-asexual, autistic person that attends Southern New Hampshire University online and is currently majoring in Communications with a focus in Public Relations. They spent the Summer of 2013 interning for Learning Games Network as a Community Outreach and Content intern. Rin currently works on PR and writing for CollectQT, and writes about MOGII pop culture for Official Gay Geeks. They spend a lot of time talking to their fiancee, and petting their cat Regina. When not writing their first novel, they can be found playing or designing games. You can read more of Rin's work at http://ckoliver.com/.

Claudina Sarahe

Claudina Sarahe (@itsmisscs) is an entrepreneur, front-end systems designer, and speaker. She resides in New York City. She co-founded and co-organizes GothamSass and SassConf. She's also a co-founding member of Quiltbag++. Claudina slings code for Oddbird and helps companies, like Casper, build performant and maintainable front-end systems. When she's not organizing code, teams, or projects, Claudina dawns her alter ego, DJ Afrocat, and spins vinyl around the city.